Reinaldo Hanoi Valdés Reinoso
Bertha Rita Castillo Edua

Proposed measures to mitigate damages

Reinaldo Hanoi Valdés Reinoso
Bertha Rita Castillo Edua

Proposed measures to mitigate damages

Due to waterlogging of olive plantations at Olive Land Farms

ScienciaScripts

Imprint

Any brand names and product names mentioned in this book are subject to trademark, brand or patent protection and are trademarks or registered trademarks of their respective holders. The use of brand names, product names, common names, trade names, product descriptions etc. even without a particular marking in this work is in no way to be construed to mean that such names may be regarded as unrestricted in respect of trademark and brand protection legislation and could thus be used by anyone.

Cover image: www.ingimage.com

This book is a translation from the original published under ISBN 978-613-9-40585-5.

Publisher:
Sciencia Scripts
is a trademark of
Dodo Books Indian Ocean Ltd. and OmniScriptum S.R.L publishing group

120 High Road, East Finchley, London, N2 9ED, United Kingdom
Str. Armeneasca 28/1, office 1, Chisinau MD-2012, Republic of Moldova, Europe
Printed at: see last page
ISBN: 978-620-7-76427-3

PROPOSAL FOR MEASURES TO MITIGATE WATERLOGGING DAMAGE IN OLIVE PLANTATIONS AT THE OLIVE LAND FARMS".

AUTHORS

DR. C. REINALDO HANOI HANOI VALDES REINOSO [1]

DR. C. BERTHA RITA CASTILLO EDUA [2]

[1]PHD IN FORESTRY SCIENCES. PROJECT 8470 COLLABORATOR. HASTINGS LA FLORIDA ORCID REYVR1806@GMAIL.COM (ORCID 0000-0003-3582-0239)

[2]PHD IN FORESTRY SCIENCES. FACULTY OF FORESTRY SCIENCES. UNIVERSITY OF PINAR DEL RÍO "HERMANOS SAIZ MONTES DE OCA". CUBA. CASTILLOBERTHARITA@GMAIL.COM (ORCID 0000-0002-8011-0175)

THINKING

2

*"Peace begins with a smile and is consolidated with an **olive branch** ... symbol of peace, wisdom and longevity".*

ACKNOWLEDGEMENTS

To all those who contributed to this achievement:

Thank you very much

DEDICATION

To our children

SUMMARY

Olive cultivation has demonstrated resistance to changes in some climatic variables, and is capable of obtaining satisfactory yields under relatively adverse soil and climatic, landscape and management conditions. The purpose of this research is to propose measures for the management of olive trees under waterlogged conditions in Olive Land Farms located in the state of Florida. A diagnosis of the area under study was carried out in order to identify the current state of the olive grove in waterlogged areas. A diagnosis was carried out to analyse the situation of the plantation taking into account the variables variety, plant condition and height. To determine the relationship between variety and plant condition, a Chi analysis was carried out^2 , and a one-factor analysis of variance was carried out for variety - plant height. Statistical differences between means were identified using Tukey's PostHoc test ($p < 0.05$). The result of the diagnosis revealed that there is an affectation caused by the period of waterlogging to which the plants were subjected and that there is no relationship between the varieties and the state of the plants, but a relationship was observed with respect to height and variety. Measures were proposed such as the creation of drainage systems, the execution of appropriate cultural work, the selection of species and varieties with greater resistance to flooding and the maintenance of a good phytosanitary state of the plantations, as well as the creation of controlled temporary flooding zones and disease control.

Key words: Olive tree, soil and climatic conditions, waterlogging, climate change.

INDEX

INTRODUCTION

Climate change is a global challenge affecting the entire planet and understanding how to adapt crops, such as olive groves, is crucial to ensure sustainability and food security.

The olive tree is cultivated in many parts of the world, of which the Mediterranean basin occupies 98% of the total area, 1.2% in the Americas, 0.4% in East Asia and a further 0.4% in Oceania (Camacho, 2019). In total, it is distributed over 11 million hectares (0.25% of total cultivated land and 25% of total permanent crops) with an average annual production of between 2.5 and 3 million tonnes. Of these, the majority is rainfed (70%) and extensive (72%), the remaining 28% is intensified in some way (intensive and super-intensive). Furthermore, the majority is for olive oil production (87%) and only 13% for table olives. The increasingly incipient area of organic olive groves stands out.

The structure of the root system varies according to the type of propagation of the material. Seedlings have a tap root in the early stages of development (Guerrero, 2003); however, in commercial plantations most trees are obtained by rooting cuttings. Depth, lateral spread and degree of branching depend on soil type and characteristics, including water content and aeration capacity. Roots demand much more oxygen, which they sometimes cannot find if the soil is waterlogged (Barranco et al., 2008).

The olive remains a species of great interest worldwide because it is the most economically important oil tree crop in temperate zones (Díez et al., 2016).In the United States, Florida may be the next agricultural region for small-scale commercial olive production, followed by California, which has more tan 30,000 acres dedicated to producing the raw material. Over the past few years, 25 commercial olive cultivars have been planted in 30 research sites and private orchards throughout Florida, where olive responses to climatic conditions are necessary to develop the local industry (Ross, 2023).

The olive tree's ability to survive prolonged periods of drought, a common situation in the Mediterranean area, is well known. However, when receives adequate water quantity and quality, fruit production yields increase considerably (Domenech, 2020).

The olive tree is a Mediterranean climate crop, characterised by a low rainfall regime that is irregularly distributed in time and intensity (Barranco et al., 2017, Lorite et al; 2019). Therefore, water scarcity is the main limiting factor for olive productivity. Paradoxically, water also causes serious damage when it occurs persistently or with strong intensity, events that occur relatively frequently, as is the case in the State of Florida, where heavy rains occur as a result of climate change, leading to intense periods of waterlogging in the region's soils.

Waterlogging is caused by floods, which are temporary overflows of water onto land that is normally dry. Flooding is the most common type of natural disaster in the United States, causing significant losses of water-soluble nutrients and hindering the uptake of some minerals from the soil. In addition, the lack of oxygen is very detrimental to the microbial (and aerobic) soil population (Salgado et al; 2019).

The same author states that the first symptom of waterlogging damage is the closure of the stomata, which eventually leads to wilting of the plant. The implicit effect of stomatal closure is that the plant stops transpiring and photosynthesis (or breathing), which prevents it from mobilising nutrients, generating them or thermoregulating. Therefore, plants subjected to hypoxia may show some or several of these symptoms:

wilting, leaf loss, reduced growth, leaf chlorosis, senescence and plant death (Salgado et al, 2019).Olive trees have been found to be sensitive to waterlogging and temperatures below -10oC (Camacho, 2019). The damage caused by flooding to crops is always proportional to the time the water remains in the cultivated plot, but crop resistance varies according to several criteria such as the species, the age of the plant, physiological state and the frequency of flooding (the greater the frequency, the greater the damage, due to the accumulation of effects). The variability of soil moisture directly affects plant growth; with low water uptake, nutrient uptake is also reduced, and the crop expresses this in a lower growth rate and hence lower yield. To improve soil conditions in the face of flood risk, actions can be carried out such as the creation of drainage systems, execution of appropriate cultural work, creation of natural protection strips next to the riverbed, reorganisation and rotation of crops, selection of species and varieties with greater resistance to flooding and maintenance of a good phytosanitary status of plantations, as well as the creation of controlled flooding areas (Salgado et al, 2019).

In Hastings, located in the state of Florida, in the period from August 2023 to January 2024, there have been numerous floods caused by heavy rains that have left the soil waterlogged, with waterlogging being one of the damages that most affect the cultivation of olive trees.

The purpose of this research is to propose measures that contribute to the mitigation of damage caused by waterlogging in olive plantations of the Olive Land Farms located in the state of Florida.

DEVELOPMENT

2.1. Impacts of Climate Change on the production of olive trees

Climate change is affecting water availability. Erratic rainfall patterns and higher temperatures can lead to droughts or floods, both of which can be devastating for agriculture. Scientific studies have documented the positive effects of olive growing on the environment. In addition to the role of olive trees in safeguarding biodiversity, soil improvement and as a barrier to desertification, there is evidence that specific agricultural practices have the capacity to increase atmospheric CO2 fixed in permanent vegetative structures (biomass) and in the soil (Granitto, 2016).Partial publications of the Sixth Report of the Intergovernmental Panel on Climate Change (IPCC) in August 2021 warn about the worsening climate situation, with recent floods in several countries around the world and the well-known episodes of torrential rains, climate change has much to do with this phenomenon that manifests itself in extreme events, such as changing rainfall patterns, rising temperatures and prolonged droughts that affect food security. Waterlogging affects agriculture by reducing soil quality and the productivity of many crops (Fisher, 2021).Predictions of changing rainfall patterns and intensity predict flooding in various parts of the world, especially in the northern Andes and more frequently at higher altitudes, while in the lowlands and southern South America flooding will be reduced. In addition, it is estimated that 13% of the land in Latin America is characterised by poor drainage due to its physiography, which is conducive to flooding. The greatest danger is for plantations near rivers, which are not only dependent on climate change but also on the occurrence of environmental factors such as heavy rainfall, storms, overflowing rivers and excessive irrigation. Incidents of waterlogging and flooding have increased in frequency and are unpredictable worldwide, especially due to erratic and unpredictable rainfall. seasonal. Rainy seasons in poorly drained soils produce anaerobic conditions that are detrimental to plant roots.Oxygen in flooded land decreases because the diffusion of gases in the water is 10,000 times slower compared to well aerated soil, which generates an energy crisis for the root tissues due to the anoxic environment, leading to the death of the plant. In addition, O2 deficiency in the soil damages microbial communities and reduces numerous oxidised nutrients (NO3-, Fe3+, SO4 2-) generating high levels of reduced compounds (Mn2+, Fe2+, NH4+, H2S) and organic compounds that can be toxic to plants.In plants, in addition to the effect on nutrient and water uptake due to energy deficiency, the most serious impact is on photosynthesis due to reduced stomatal conductance and stomata closure, as well as reduced leaf growth, chlorosis, leaf scorch and finally leaf drop. Excessive soil moisture conditions favour the incidence of pathogens. It is also important to take into account that high soil and/or water temperature and high solar radiation during waterlogging increase their adverse effect on plants (figure 1).

Figure 1 Waterlogged land at Olive Land Farms

Source: Own elaboration

Waterlogging caused by floods can affect agriculture, reducing soil quality and the productivity of many crops. In the specific case of olive crops, this phenomenon can have negative consequences. For example, at Olive Land Farms, located in the state of Florida, USA, damage has been observed due to the waterlogging to which the olive plants have been exposed. Extreme events due to waterlogging of agricultural land will continue to increase, also in places where it was not expected before, so a multi-faceted approach is needed, including the breeding of tolerant varieties and rootstocks, studies of the physiology of the waterlogged plant and appropriate management measures to cope with this climatic risk such as land drainage.

2.2. Drainage of waterlogged land

Soil drainage is a very important technique for maintaining plant health and preventing water accumulation in unwanted areas. Drains are systems that allow water to flow out of the soil, thus avoiding soil saturation. There are several types of drains and their installation depends on the type of soil and the use of the area.

Types of drainage

Natural drainage is the most common and occurs naturally when water seeps through the soil and evaporates. However, in some areas, the soil may be too wet and additional drainage techniques are needed.

Underground drainage is one of the most effective methods of draining the soil. A network of underground pipes is used to collect the water that accumulates in the soil and transport it away from the area. This type of drainage is very useful in areas with clay or compacted soils. Surface drainage is another popular method. This type of drainage uses channels or ditches to collect water that accumulates on the soil surface and transport it away from the area. It is especially effective in areas with sandy soils or steep slopes.

To drain waterlogged soils and keep them healthy, it is important to follow a number of key steps:

1. Identify the cause of waterlogging:

Before starting any drainage work, it is essential to identify why waterlogging is occurring in the ground. It may be due to heavy rainfall, bad weather

natural drainage, clay soils, among other causes. This information will help you determine the best solution.

2. Improve soil structure:

If the problem is due to compacted or clay soils, it is advisable to work on improving the soil structure. This can be achieved by adding organic matter, such as compost or peat, which will help to improve water infiltration and reduce waterlogging.

3. Install drainage systems:

For more severe cases, it may be necessary to install drainage systems, such as drainage ditches, perforated pipes or soakaways. These systems will help to divert excess water and keep the soil in a healthy condition for plants.

4. Choose plants that are tolerant to excess water:

Once you have drained the soil, it is important to choose plants that are tolerant of excess water, such as water lilies, iris or water hyacinths. These plants will not only survive in wet soil conditions, but will also help to absorb excess water and maintain the balance of the garden.

Identification of the cause of water accumulation on the ground

Identifying the cause of the water accumulation is the first crucial step in solving the problem of waterlogged soils and determining the reason behind this accumulation will enable the right solutions to be applied effectively.

Some possible causes of water accumulation on the ground include:

Poor soil drainage: When the soil is compacted or clayey, water tends to accumulate on the surface instead of infiltrating properly.

Topography of the terrain: If your garden is on a slope or in a low-lying area, water is likely to accumulate in certain areas.

Clogged drainage systems: Blocked or poorly maintained drainage pipes can lead to water accumulation in the garden.

Once the specific cause of the water accumulation has been identified, the most appropriate drainage strategies can be selected to effectively solve the problem.

Strategies to improve drainage on terrain

Installation of subsurface drains: Allows for the collection and diversion of water to designated drainage areas.

Creation of drainage ditches: Helps to redirect water away from waterlogged areas to places where it can be absorbed by the soil.

Soil improvement: Add organic matter or sand to the soil to improve its drainage capacity.

Selection of tools for draining waterlogged land

Selecting the best tools for draining waterlogged land is essential to ensure that the drainage process is effective and efficient. Having the right tools will make the job easier and help to achieve optimal results in improving the health of the soil.

1. Shovels and hoes:

These tools are essential for digging trenches and opening drainage channels in waterlogged soil. A shovel with a sharp point will facilitate the task of removing the soil and creating the necessary channels for the drainage of accumulated water.

2. Bubble level:

Using a spirit level will help ensure that the slopes you are creating in the ground are adequate to direct water to the drainage areas. Maintaining an even slope in trenches is key to effective drainage.

3. Drainage pipes:

Drainage pipes are essential for channelling water away from waterlogged areas. You can opt for perforated pipes that allow water to seep slowly into the surrounding soil or smooth pipes for faster and more direct drainage.

4. Gravel or drainage stones:

Placing a layer of gravel or stones in the bottom of drainage ditches will help prevent clogging of pipes and facilitate the flow of water through the drainage system. The gravel also helps prevent soil erosion in the drainage areas.

5. Ground leveller:

Having a soil leveller will allow you to ensure that the soil surface is even after the drainage process has been completed. This is important to avoid the formation of new waterlogged areas in your garden.

When choosing tools to drain waterlogged soil, it is important to consider the extent of the waterlogging problem, the size of your garden and your gardening skills. If in doubt, it is always advisable to consult a professional for specific guidance.

Implementation of an effective drainage system.

Implementing an effective drainage system is essential to avoid problems such as waterlogging that can damage plants and soil. Below are the key steps to achieving optimal drainage and keeping the soil healthy:

1. Identify problem areas

Before starting the installation of the drainage system, it is essential to identify areas that tend to waterlog. This can be done by observing where water collects after heavy rain or when watering plants. Marking these areas will help you plan the location of drains effectively.

2. Choosing the right type of drainage

There are different types of drainage systems, such as French drains, drains and infiltration trenches. It is important to select the type that best suits the needs of your garden and the level of waterlogging you have. For example, French drains are ideal for areas with surface water accumulation problems.

3. **Prepare the ground**

Before installing the drainage system, it is crucial to prepare the ground properly. This may involve digging trenches, drilling pipes or installing drains depending on the type of drainage chosen. Make sure you have the necessary tools and follow the installation instructions to the letter.

4. **Install the drainage system**

Once the ground is ready, proceed to install the drainage system following the previously planned design. It is important to ensure that the pipes are correctly laid and that the drains drain the water efficiently. A good drainage system will ensure that excess water is drained away from the roots of the plants, preventing damage due to waterlogging.

5. **Carry out tests and adjustments**

After installation, it is advisable to carry out tests to ensure that the drainage system is working properly. Simulate rainfall or water the olive grove and observe how the water flows through the system. Make any necessary adjustments to improve drainage efficiency and ensure that all problem areas are covered.

2.3. The Olive Tree. Its characteristics

Figure 2 Olive tree

Source: aceitedeoliva.com

Olea europea L (olive tree) is a tree species belonging to the botanical family Oleaceae, distributed in tropical and temperate regions. It is native to theIt is the only edible oleaceae and has been cultivated in the Mediterranean for more than 6,000 years. It is made up of 29 genera and around 600 species spread almost all over the world. Within this large family we can find such well-known species as ash (Fraxinus angustifolia), jasmine (Jaminum officinalis), lilac (Syringa vulgaris), privet (Ligustrum sp.), and others of the genus Forsythia and Osmuanthus.1

The genus Olea is composed of more than 30 species, all of which come from areas with relatively difficult growing conditions (Zohary, 1973). Most are shrubs or trees. The only species with edible fruit is Olea euopaea to which the olive belongs. There is controversy about how to subclassify the species, but cultivated olive trees are generally considered to belong to the subspecies sativa and wild olive trees to the subspecies sylvestris (Barranco-Navero et al., 2017).

Botanical sketch:

Family: Oleaceae.

Genus: Olea.

Species: Olea europaea. Olea europaea var. sylvestris (wild olive tree)

Olea europaea var. sativa (cultivated olive tree)2

Olea europaea var. sativa or Olea sylvetris is distributed in Spain, Portugal, North Africa, Sicily, Crimea, Caucasus, Armenia and Syria. The subspecies Laperrini occurs in North Africa from the Moroccan Atlas Mountains to Libya, and is found spontaneous even at an altitude of 2,700 m.

In the latest studies of olive tree varieties in Spain carried out by the Agronomy Department of the University of Cordoba (1972-1992), 262 varieties or cultivars of olive trees have been identified in Spain, of which 24 varieties predominate and are the best known.

"This diversity is probably due to the autochthonous origin of the varieties, which led to the choice of different cultivars in each area, and to certain factors which have maintained the initial morphogenetic situation. Genetic homogeneity within the cultivated varieties is very marked due to the vegetative propagation procedures used.

2.4. Characteristics of the olive tree

The olive tree is an evergreen tree that in the right conditions can reach up to 15 metres in height. In fact, the sinuous trunk with its dark, rough bark can reach a radius of more than 100 cm in adult plants.

The stem is a short trunk which then branches out irregularly to form a very tight crown. The trunk has particular protuberances due to its permanent lateral growth and bark of grey-greenish tones.

The plant ensures anchorage through a strong main root. It has a group of absorption roots that guarantee the absorption of water and nutrients.

The branching of the olive tree is organised into first, second and third order branches. The trunk and the first order branches form the main structure, the secondary branches, which are less voluminous, support the tertiary branches, where the fruit develops.

The simple, persistent, lanceolate or elliptical leaves with straight margins are leathery in consistency and bright green in colour. On the underside, the colour is greyish, with abundant trichomes whose function is to control the circulation of water and filter the light.

The yellowish-white flowers consist of a calyx with four persistent cup-shaped sepals, united at the base. The corolla has four creamy-white, mutually concrescent petals and two short stamens with two yellow anthers.

The inflorescences are grouped in clusters that arise from the leaf axils, containing between 10-40 flowers on a central rachis. The fruit is a globose drupe of 1-4 cm green that turns black, reddish or purplish when ripe. The fruit (the olive) contains a single large seed. This olive is characterised by a fleshy, oleaginous, edible pericarp and a thick, rough, hard endocarp.

Six naturally occurring subspecies of Olea europaea have been described with a wide geographical distribution:

West Africa and south-east China: Olea europaea subsp. cuspidata.

Algeria, Sudan, Niger: Olea europaea subsp. laperrinei.

Canary Islands: Olea europaea subsp. guanchica.

Mediterranean basin: Olea europaea subsp. europaea. Madeira: Olea europaea subsp. cerasiformis (tetraploid). Morocco: Olea europaea subsp. maroccana (hexaploid).

2.5. Etymology

The word "olive" comes from the Latin "olfvum", which refers to the same tree. In turn, this Latin term is a loan from the Greek "éÀ mov", which has different meanings depending on its gender. For example, the neuter form of "éÀ mov" refers to the product of olive oil. Furthermore, since ancient times, the olive tree has been considered a symbol of peace, and in popular etymology, the Greek word "elaion" (oil) has been related to "eleos" (mercy).The olive tree has a rich history and meaning in different cultures.

2.6. Habitat and distribution of olivo

The olive tree is native to the southern Caucasus, the high plateaus of Mesopotamia, Persia and Palestine, including the Syrian coast, and of course the Mediterranean basin (Spain, Italy, Greece).It was first cultivated around 7,000 years ago throughout the Mediterranean, and in 3000 BC it was grown on the Greek island of Crete, and it is believed that the olive tree was the main source of its wealth. Spanish missionaries introduced the crop to the Americas in the mid-16th century, initially in the Caribbean and Mexico. Subsequently, it was dispersed in North America (California) and South America (Colombia, Peru, Brazil, Chile and Argentina). This plant grows in a range of 30-45° north latitude and south latitude. Particularly in climatic regions with hot, dry summers and where winter temperatures do not fall below freezing.

2.7. Properties of olives/olives

The fruit of the olive tree, called olive, is a simple, fleshy berry, globular or ovate in shape depending on the variety, 1-3 cm in size. When tender they are green in colour and when ripe they turn blackish or dark green, with thick flesh.The pulp, or thick, fleshy and oleaginous sarcocarp, is edible, and the endocarp containing the seed is bony and firm. Olives require a curing and maceration process to be consumed, either directly or as a garnish in various gastronomic specialities. Olive oil, a monounsaturated fat with a high oleic acid content, is extracted from olives. Olive oil is beneficial for the health of the cardiovascular system by regulating cholesterol.

Olive oil has digestive properties, has a laxative, diuretic, astringent, cholagogue, emollient, antiseptic, hypotensive and anti-inflammatory effect. It is also used to soothe burns, insect bites, sprains and strains, and to heal conditions of the mucous membranes.

2.8. **Uses of the olive tree** .

The leaves, fruit and oil of this plant have been used for thousands of years in gastronomy, traditional medicine and decoration, to the point of being considered the most valuable crop in the Mediterranean region. The leaves are used as an antiseptic, astringent and to soothe fever and skin scrapes, although it is also believed to be a remedy for hypertension. The fruits need to be macerated before consumption, but some varieties can be eaten after drying. They are eaten as an appetizer or as an accompaniment to many dishes, as they are a staple ingredient of Mediterranean cuisine. Olives contain about 40 percent or more oil, which is extracted after harvesting and used in cooking, as a natural remedy and as a cosmetic treatment. It provides flavour and texture to salads, sauces and mayonnaise, as well as to pizzas and pastas.

It is also used in the treatment of constipation, peptic ulcers, burns, itching and dandruff, among others. As it is rich in oleic acid, it is believed to be beneficial for cardiovascular health. Moreover, the wood of the tree is useful as fuel and raw material for furniture and construction, and the tree can be planted as an ornamental motif.

2.9. Diseases of the olive tree and their treatment natural

Although there are many diseases that can affect olive trees, there are a few diseases that are the most important in terms of damage and number of attacks. In addition, there are also a large number of pests that cause losses, such as the olive milkweed, although these are not considered as diseases as such.

Root asphyxia of the olive tree

Excessive moisture in the soil is not beneficial for the olive tree. Root asphyxia can cause weakening, chlorosis, yellow leaves, olive drop, defoliation and fungal growth on the trunk of the olive tree.

Repilo

Scientific name Cycloconium oleaginea, it is one of the worst diseases that can damage an olive tree crop. This disease causes the leaves of olive trees to fall off, as well as a significant reduction in fruit production and, in the long term, a general weakening of the tree. The most visible symptom is the appearance of circular spots on the leaves, which can sometimes also be seen on the fruit. The leaves eventually take on a whitish colour and drop prematurely.

There are no varieties of olive tree that are completely resistant to this fungus, and once the pest has appeared, there is nothing to do but apply fungicides, which must be used in autumn and at the end of winter. It is important to concentrate the fungicide application on the lower part of the canopy.

Verticillosis

Verticillium is another very dangerous disease for olive trees, and cases of it have increased greatly in recent seasons. It is quite difficult to combat and is present throughout the Mediterranean basin, with particular incidence in Andalusia.

The fungus enters the plant through wounds or through its roots and the first symptom it produces is discolouration of the leaves, which curl in on themselves around the central nerve. The flowers and fruit eventually dry out as well, and it can even kill the tree. It is very dangerous because it spreads through a large number of plants and can survive for more than a decade in the soil. In combating it, it is very important not to spread the infection by working the soil and fertilising. Moreover, there are varieties that are resistant to it. Otherwise, only methods such as solarisation or anti-fungal substances can be used, as well as pruning and control of the infected parts.

Anthracnose on olive trees

Soap olive or anthracnose is currently considered the most important disease affecting olive trees worldwide. It causes a great deal of damage to the crop, spoiling the oil of the affected fruit.

Symptoms are olives with brownish, wrinkled spots and leaves with dry, necrotic parts, which eventually dry out completely. Branches also dry out at the tips and inflorescences may also be affected.

In the case of anthracnose, resistant varieties such as picual can be chosen, the tree should be well ventilated between the branches and fungicides should be applied. The olive tree requires relatively little care, provided it is planted in a soil that meets its minimum requirements. It is a species that adapts to low fertility and sandy soils, although it requires sufficient sunlight.

It does not tolerate prolonged cold, as defoliation of tender leaves and abortion of flower buds may occur. Young plants are more prone to strong winds than adults, so they need windbreaks in exposed areas.

The olive tree grows and develops well in maritime areas, however, it is susceptible to high levels of soil salinity. Although sensitive to frost, it requires a low temperature level to maintain flowering and increase production.

Irrigation should be continuous in the establishment stages of the crop, and in productive plants hydration increases productivity. Excess nitrogen fertiliser increases leaf area production and crown weight, which can lead to tipping. It is recommended to place a layer or organic mulch around the stem in order to maintain humidity and control weeds. Maintenance pruning is recommended, leaving three to five branches to facilitate light and water penetration (Lidefer, 2023).

2.9. Olive varieties grown at "Olive Land Farms".

At Olive Land Farms, eight varieties of olive are grown: Arbequina, Arbosana, Ascolana, Kalamata, Koroneiki, Picual, Manzanilla and Taggiasca (figure 3).

Figure 3 Varieties of olive grown at Olive

Land Farms

2.9.1. Arbequina

Arbequina is a variety of olive tree (Olea europea) that was introduced by the Duchess of Cardona in the 17th century, who lived in the castle-palace of Arbeca, Catalonia, hence the name given to this variety in honour of the municipality where she lived (Figure 4).

Figure 4 Arbequina.

Features

It is a small tree. A size that is deceptive, since at first sight it seems that not much juice can be extracted from its interior, but this is not the case. Why? Because it is one of the fattest varieties. However, it has a large

26

stone. This means that its pulp to stone ratio is low. The arbequina is characterised by a great resistance to cold, a very low vigour and a low resistance to calcareous soils. The size of its fruit is the smallest of the varieties cultivated in Spain, between one and two grams.As for its distribution, we can find plantations in the communities of Catalonia, Aragon and Andalusia in Spain, in the area of Maule in Chile, La Rioja in Argentina, Minas Gerais in Brazil and recently in areas of the sierras of Maldonado and Minas in Uruguay. It is the basis of modern intensive plantations as its low vigour allows a high density of plants, reaching 2000 plants per hectare in some plantations.Early arbequina extra virgin olive oil is mild and has a fruity aroma. However, due to the low proportion of polyphenols, its shelf life is limited. Another characteristic that makes it unmistakable is its appearance. It is a symmetrical and spherical olive. The arbequina olive reaches maturity when its skin is black. However, as they are not exposed to oxidation, they are harvested early. This means that they are harvested when they are still a little green. This guarantees a fruity flavour.

Uses

In addition to being used as a table olive, it is also used to produce high quality olive oil. In general, olive oils based on this type of olive tree are more fluid and sweet, with a finish where bitterness and spiciness are barely perceptible (Corral, 2020).

2.9.2. Arbosana

Olea europea var. Arbosana (Figure 5)

Figure 5 Olea europea var. Arbosana

Features

It is one of the most popular varieties of olive trees today. It is native to the Penedés region, located between Barcelona and Tarragona, in Spain. Due to its reduced vigour and high yield, it has become one of the favourite varieties for hedge planting. Unlike other varieties, they are not very resistant to cold and tolerate periods of drought rather poorly (Aparicio, 2023).

These olives are usually planted in hedgerows, due to their low vigour and high productivity. As a result, the density of this type of olive grove is higher than that of other oil varieties. Arbosana harvests are three weeks later than Arbequina, so a wonderful option is to combine both varieties on the same land. This type of olive trees are very regular in terms of productivity, so they tend to produce the same amount of oil season after season and they have very few alternate harvests.

Due to its production level of more than 2000 kg of oil per hectare and its unmistakable flavour, many areas of Spain and other countries have opted for this variety of olive tree, which is why the expansion of arbosana oil is increasing all the time. Moreover, its productivity is quite precocious, as it achieves an optimum yield in a short period of time. If there is one thing that characterises Arbosana olive trees, it is their low vigour and their good adaptation to super-intensive plantations. This is due to the fact that this variety of olive is very difficult to detach.

Arbosana olive trees stand out for their reduced vigour, which makes them ideal for super-intensive plantations, and within two years of planting, they can already obtain interesting harvests. With proper care, they can produce more than 2 tons of olive oil per hectare, have an excellent rooting and expansion capacity, unlike other varieties of olive trees, Arbosana has a later flowering, resists well to repilo and verticillium, but is somewhat susceptible to cold and tuberculosis of the olive tree.Supremo Arbosana is an early harvest oil, cold extracted by mechanical processes.

As we have mentioned before, this is a variety that does not usually occur in Andalusia. However, this oil is produced in Jaén and, being surrounded by Picual olive trees, it obtains very interesting nuances. For this reason, its production is quite exclusive, around 3,000 bottles per harvest.

Uses

Arbosana is grown to produce olive oil.

The nose is medium fruity with herbal and fruity aromatic notes. However, it has a persistent flavour on the palate, although its entry at first is quite smooth. It is ideal for those who prefer sweet flavours, but with character; those that, despite their smoothness, leave an indelible mark in the memory of the diners who try it.

2.9.3. Ascolana

Olea europaea var. Ascolana (Figure 6)

Figure 6 Olea europaea var. Ascolana

The **olive** cultivar **Ascolana Tenera**, also known as Ascolano, has its origin in the Italian territory of Ascoli Piceno, in the Italian region of Marche (Marche). The main olive groves are in the Marche region, with little spread in central Italy. Its main use is the production of olives for table consumption (olives 'alla ascolana').The variety needs loose and fresh soils of limestone composition to reach its productive potential, the Ascolana olive tree is vigorous, with erect growth habit **and** high crown thickness, the Ascolana leaf is elliptic-lanceolate in shape **and** medium size, the Ascolana Tenera olive tree variety is resistant to cold, cold, limestone soils, tuberculosis and cochineal.It is particularly sensitive to fly attack, which is attracted by its large size, the plant has good rooting capacity, it is an early ripening olive variety, **it** has very large olives (average 7 grams). Some fruits can reach 10 grams, thanks to their size, Ascolana olives are well known internationally. When ripe, Ascolana olives reach a black colour **and** produce olives with a medium yield (17%). Sensitive to the fly. Its production decreases if the soil is not suitable. It is also resistant to several olive tree diseases.

Uses: Ascolana olives are good for dressing and are a highly valued variety of olive for eating 'olives a la ascolana'.

2.9.4. Kalamata

Olea europaea var. Kalamata (Figure 7)

Figure 7 Olea europaea var. Kalamata

Features

The tree is quite robust, its branches have a climbing tendency and large leaves. On average, the fruit weighs 5-6 grams, the kernel is smooth and detaches easily from the flesh. This variety produces vigorous, upright, moderately cold-resistant olive trees, with late, abundant production and large fruit. Ideal for canned black olives and oil. Kalamata olives are so called because they were originally grown in the region around Kalamata, which includes Messenia and nearby Laconia, both located on the Peloponnese peninsula. They are now grown in many parts of the world, including the United States and Australia. They are dark purple, plump and almond-shaped, from a tree that is distinguished from other olive varieties by the size of its larger leaves (Antol, 2004). The trees are cold intolerant and susceptible to verticillium but are resistant to Pseudomonas savastanoi and olive fruit fly.They cannot be harvested and must be picked by hand to avoid bruising. They are classified as black olives.

Uses

They can be eaten directly as an appetizer or as part of other dishes, such as the typical Greek salad and other types of salads as well as in the preparation of cooked dishes where olives are used as an ingredient.

2.9.5. Koroneiki

Olea europaea var. Koroneiki (Figure 8)

Figure 8 Olea europaea var. Koroneiki.

Features

It is a tree of medium vigour, open growth habit and dense crown. It has a pointed apex and a rounded base, with small lenticels. The bone is small, elongated and asymmetrical. It has very particular elliptical leaves, whose main characteristic is that they are shorter and narrower, easily distinguishable from other varieties of olive tree. This is the foliar adaptation strategy, which allows it to withstand climatic conditions in arid areas. It has become widespread thanks to its adaptation to super-intensive plantations. It is a less productive olive tree than Arbequina, its greater vigour can be a problem in adapting to the super-intensive framework in fertile soils and with good weather conditions. It is a variety with a very early entry into production, with high productivity, but somewhat lower than Arbequina.The Koroneiki variety of the olive species is very resistant to arid soils, although it is not very tolerant to low temperatures. For this reason, it has spread successfully along the Mediterranean coasts and is not recommended above 400 m altitude. It should also be noted that they are cold-resistant trees, adapted to hedgerow or super-intensive plantations and with little alternate bearing, i.e. they have a constant production over the years. Although it is quite productive, it is not among the most productive olive trees. This olive variety is early maturing about two weeks after arbequina and has a good yield, in some areas it reaches yields between 18-20%. One of the highest yielding olives.

Uses

The fruit is more oval than normal and has little pulp, which is why it is mainly used for oil production and not for commercial olives.

2.9.6. Camomile

Olea europaea var. Manzanilla (Figure 9)

Figure 9 Olea europaea var. Manzanilla

Features

The Manzanilla Sevillana olive tree is a table olive variety par excellence. It is grown both in Spain, where it occupies an area of around 100,000 hectares, and in other countries such as Argentina, Australia, Israel, Portugal and the United States, where the Manzanilla de Sevilla olive variety is known as Manzanillo Olive.It is a tree with low vigour, open growth habit and medium crown density. Its fruit is heavy, spherical and symmetrical. Its apex is rounded and the base is truncated. The stone is heavy, ovoid and symmetrical. It is of reduced vigour, early entry into production, high and alternate productivity.Its cultivation requirements are somewhat demanding, especially outside its native area. It is sensitive to cold, root asphyxia and iron chlorosis, verticillium and tuberculosis.

Uses

It is the most widespread variety of olive for the table. The excellent characteristics of its flesh and the ease of extraction of the stone make it the main variety for dressing, both green and black. It is the most appreciated table variety, it is also used for oil but it has a low yield, however it is of high quality and stability.

2.9.7. Olive tree Picual

Olea europaea var. Olivo Picual (Figure 10)

Figure 10 Olea europaea var. Picual

Features

It is a tree of medium vigour, open growth habit and thick crown, with a medium weight, ovoid and asymmetrical. Rounded apex and truncated base. Abundant lenticels. The fruit is black when ripe, the stone is heavy, elliptical and asymmetrical with a rough surface.

It has a high production, with little alternation when well cultivated. It is early maturing and with low resistance to fruit detachment. It is very hardy and therefore adapts to many types of soils and climates, resistant to cold and salinity. It is a variety tolerant to tuberculosis, but very sensitive to whorls. In areas susceptible to whorls it is very important to analyse the soil before planting.

The picual olive tree is widely planted due to its high fat yield, early ripening, ease of cultivation and high oil quality. The picual variety tree is robust, with short branches and adapts to any type of soil, although it is sensitive to periods of drought. The Picual olive tree, the most important variety in Spain, has excellent characteristics to produce olive oil (high yield olive, high productive capacity, easy release). It is the type of olive tree (Olea Europea Sativa), most present in the Andalusian olive grove and is the most cultivated olive in Spain. It has an increasing cultivation area and currently accounts for approximately one million hectares of olive groves. The main growing area is Jaén, where more than 90% of the olive trees planted belong to the Picual olive variety. In the Andalusian provinces of Cordoba, Granada and Seville, there is also an important extension of the crop.

Its cultivation area continues to expand, although it does not adapt to hedgerow olive groves, it is regular, productive and easier to harvest than other varieties.

Its most probable origin is in Spain, however, they are very old varieties that do not allow us to be sure of their century or their exact area of origin.

This olive tree is also known by other names: Nevadillo, Nevadillo Blanco, Marteño, Lopereño, Corriente, Andaluza, Picúa, Blanco, Fina, Morcona, Jabata, Nevado, Nevado blanco, Nevado Lopereño, Salgar and Temprana.

The name "Picual" is due to the beaked shape of this variety of olive tree. The variety is also popularly known by the name Nevadillo Blanco, this is because its leaf is whitish compared to the Nevadillo Negro variety of olive tree.

It is a variety of olive tree with excellent characteristics. It has been and still is the most profitable olive variety for traditional olive growing. It is a variety that enters into production early and is highly productive. With support irrigation and appropriate pruning, it takes about 5-6 years to produce good harvests.It has a good floral induction, an aspect that makes it a variety of olive that is not very neighbourly. It can alternate high harvests with medium harvests or good harvests if the care of the Marteño olive tree is adequate.

Picual tolerates tuberculosis well and is less palatable to olive fruit flies than other varieties. However, its leaf is sensitive to the repilo fungus and is attacked by Prays and Cochineal pests. It can be easily propagated by cutting or in nurseries by misting (seedling staking). This is why olive tree nurseries can sell Picual seedlings at a good price.

It is early ripening, the fruit is medium-sized, the olive produced by the Nevadillo Blanco olive tree has an elliptical-asymmetrical shape and reaches maturity when the skin is black, it has a medium-high pulp/stone ratio, the stalk of the Picual olive is very small.

Picual olives have a high oil yield. The oil yield will depend on the nutritional state of the olive tree, the level of load, the ripeness of the olives and the olive, climatic conditions of the growing area. The fat yield of the Picual olive is usually over 20%.

It has a low resistance to detachment, an aspect that facilitates harvesting. Even so, it holds up well on the olive tree, without falling to the ground easily.

It is a productive variety, with little influence from the ageing of the olive tree. Picual oil is stable and has good characteristics. Easy to harvest and multiply vegetatively.

Uses:

The Picual olive variety is used for oil production. Its quality for table olives is much lower than that of varieties such as Manzanilla Cacereña.

2.9.8. Taggiasca

Olea europaea var. Taggiasca (Figure 11)

Figure 11

Olea europaea var. Taggiasca

Characteristics:

Large trees, high vigour, open-grown, with a medium dense canopy. Adapted to more coastal or higher areas. It has elliptic-lanceolate leaves and medium size, it is sensitive to cold and drought. The plant has a low rooting capacity. It adapts well to different conditions, both close to the sea and in higher altitude areas. It stands out for its good productive characteristics and the high valuation of the oil and olives produced, it is of early entry into production and of high production, it obtains regular productions. It has medium flowering and its pollen is partially self-compatible. The Taggiasca olive tree (also known as Taggiasche olive) is a variety of olive tree very appreciated for its olives and its oil, its vigour is high, and its bearing is open-weeping, with a medium dense crown. The leaves are elliptic-lanceolate and of medium size. This variety enters into production early and has a high yield. It produces regular yields, flowers at an average time and its pollen is partially self-compatible. It produces a high quantity of pollen and has low ovarian abortion.Despite their small size (less than 2 grams), Taggiasca olives are appreciated for their good taste. They are elliptical, symmetrical and have no pit. The olives ripen late and turn brown-black when ripe.

Uses

The Taggiasca olive variety is used both for oil production (high yield and high productive capacity) and for table olives, where it is appreciated for its excellent flavour.

METHODOLOGY

The research was carried out at Olive Land Farms in northern Florida, located in the small town of Hastings 32145 (figure 12). It is part of the Southern Region of the United States, bordered to the west by the Gulf of Mexico and Alabama, to the north by Alabama and Georgia, to the east by the Atlantic Ocean and to the south by the Straits of Florida.

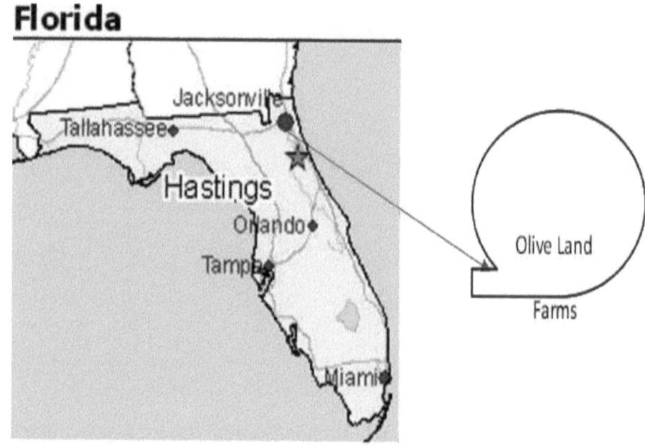

Figure 12 Geographical location of the study area

Edaphoclimatic characteristics

Florida is morphologically a vast plain, covered by calcareous sedimentary soils, located a few metres above sea level and characterised by poor drainage that favours the development of marshy areas. It is provided with lakes of karst origin, including Lake Okeechobee, the second largest lake in the United States, located four metres above sea level (m.a.s.l.) and 4-5 m deep. The climate is tropical, with two seasons of equal length: the dry season (December to April) with little rainfall and high temperatures, and the rainy season (May to November) with rainfall in the order of 1000-1650 mm of rain per year, characterised by frequent hurricanes in the late summer to early winter period. Average maximum temperatures in July are normally around 32-34 °C. The average minimum temperatures in January range from 4-7 °C in the north from Florida to over 16 °C from Miami southwards. With an average daily temperature of 21.5 °C, it is the warmest state in the United States.

Precipitation is most commonly in the form of rain, often torrential. It is a very storm-prone state and vulnerable to hurricanes entering from the Caribbean Sea from June to November (United States National Arboretum, SA).Florida's flat landscape is covered by a network of more than 1,700 streams and tens of thousands of lakes (mostly in the central region). The soils are clayey with a tendency to retain moisture and the subsoil is rich in nutrients, making them useful for agricultural purposes, although they can deteriorate rapidly when eroded (Terrasa, 2019).

3.1. Methodology used

For the development of the research, a diagnosis was carried out with the purpose of analysing the current state of the olive plantation and its affectation by the waterlogging conditions of the land, where the behaviour of the crop and its adaptation to the different situations were analysed. Eight varieties of olive trees were planted in a total area of 1.15 ha with a planting frame of 3 x 3 m and data were collected on variety and plant height. In addition, measures were proposed to contribute to the mitigation of waterlogging damage based on the methodology presented by Savé (2016). A completely randomised design was used. The relationship between variety and plant condition (healthy and with yellowing or root asphyxia) was determined, and a contingency table analysis and Pearson's Chi-square statistical significance test were applied to determine whether there are statistically significant differences between the two variables. In addition, a one-factor analysis of variance was performed to compare the mean of the height groups with respect to the variety, using Tukey's post hoc test ($p<0.05$). Statistical analysis of the data was processed in SPSS software ver. 23 for Windows.

RESULTS AND DISCUSSION

The diagnosis carried out corroborated that the level of waterlogging in the farm areas and the lack of soil aeration are affecting the development of the olive crop (figures 13 and 14).

Figure 13 Waterlogged area

Figure 14 Plantation of Olea europeae var. Arbequina in waterlogged soils.

During the tour of the area, the main manifestations detected in the plants were the presence of dead lower leaves, plants with stagnation or reduced growth and lack of vigour, root asphyxia and plants with yellowing leaves (figure 15).

Figure 15 Plantation of Olea europeae var. Arbequina with yellowing

Similar results were obtained by Arquero and Serrano (2020) who corroborated that prolonged conditions of waterlogging or high soil moisture content can cause damage due to root asphyxia (deficient oxygenation of the roots), and can even lead to the death of the tree.

Soil problems are generally related to water shortage, however, an excess of moisture can drastically damage crops. This usually happens in olive cultivation during the rainy season, water accumulates in the plots and if the soil is not well drained, anoxia (lack of oxygen) conditions occur in the root system and in a matter of days the leaves turn yellow and their growth is notoriously reduced.

These results coincide with those of García-Ruiz et al. (2020), who state that in recent years new olive plantations have proliferated on flat soils with a high clay content, where root asphyxia damage occurs after periods of heavy rainfall. As the root system is less developed, mortality due to asphyxia is higher in young trees.

Also, the structure of the root system varies according to the type of propagation of the material. Seedlings have a tap root in the early stages of development (Guerrero, 2003); however, in commercial plantations most trees are obtained by rooting cuttings. Depth, lateral spread and degree of branching depend on soil type and characteristics, including water content and aeration capacity (Barranco et al., 2008). Excess moisture in the soil, far from producing benefits, can end up suffocating the roots of the olive tree. These roots can be a source of rot where fungi develop easily.

The sensitivity of olive trees to waterlogging depends on the variety, although tolerance is generally low. However, the olive tree can recover quickly if waterlogging is temporary (Figure 16).

Figure 16 Plantation of Olea europeae var. Arbequina

Analysis of the influence of waterlogging on variety.

In the statistical evaluation of the results according to the analysis of the contingency table and the Pearson's Chi-square statistical significance test, it was shown that there is no relationship between the varieties and the state of the plant (healthy and with yellowing). Therefore, it is inferred that the condition of the plants is directly influenced by the waterlogging conditions of the soil (medium, low and prolonged), however, it was found that the prolonged period of waterlogging has a negative influence on the development of some of the olive varieties. The varieties most affected by waterlogging conditions were Arbequina and Arbosana. Excess moisture in the soil, far from producing benefits, can end up suffocating the roots of the olive tree. These roots can be a source of rot where fungi develop easily. The olive tree's sensitivity to waterlogging depends on the variety, although as a general rule its tolerance is low. However, the olive tree can recover quickly if the waterlogging is temporary. In this sense, resistance to moisture in olive trees is influenced by several factors, including climate and the genetic characteristics of the plant. These factors interact with each other and can determine the ability of an olive tree to survive and thrive in wet conditions. Temperature extremes and excess rainfall are important challenges to consider when selecting varieties.

Likewise, the qualitative results obtained are similar to those of Bernal Martínez (2017), who highlights that the variety's traits are highly heritable and conserved, and therefore little affected by environmental conditions and agronomic management.These results differ from those obtained by Moraima and Ferreres, (2002), in a study with Albizia lebbeck, Lysiloma latisiliquum and P. piscipula during two months of waterlogging where no significant differences in seedling height were shown). A period in which a good water status should be especially ensured is from bud burst to flowering, as this is when maximum shoot growth takes place, and vegetative growth is one of the parameters most sensitive to water stress.Both the level of hydration in spring and the rate at which water stress develops are decisive for the vegetative growth, development and physiology of the crop. The one-factor analysis of variance allowed comparing the mean of the height groups with respect to the variety, with Tukey's post hoc test ($p < 0.05$ (Table 1).

Table 1. One-factor analysis of variance (variety-height)

Tukeya HSD groups	Average	Variance	
V1 Arbequina	1.227	0.005	**
Groups	Average	Variance	HSD Tukeya
V1 Arbequina	1.227	0.005	**
V2 Arbosana	0.995	0.024	NS
V3 Ascoian	1.065	0.049	NS
V4 Kaiamata	1.0 2	0.014	NS
V5 Koroneiki	1.029	0.017	NS
V6 Manzaniiia	0.914	0.090	**
V7 Picuai	1.05	0.045	NS
V Taggiasca	0.941	0.023	**

The results show that the Arbequina variety is the one that shows the greatest significant difference among all the varieties investigated. Similar results are supported by Balam (2023) when he states that the Arbequina variety is recognised for its high resistance and productivity, it is a plant that has excellent resistance to cold, salinity and diseases such as repilo and tuberculosis.It can be sensitive to verticillium, and to chlorosis in very chalky soils, but in general it has a fairly interesting hardiness and characteristics that allow it to adapt to multiple soil-climate circumstances.

Table 2. Summary of the one-factor analysis of variance, Tukey's post hoc test ($p < 0.05$).Height

Origin from variations	Sum of squares	Degrees of freedom	Mean squares	F	Probability
Between groups	0.6524	7	0.093	2.776	0.013
Inside of the groups	2.4171	72	0.034	-	-
Total	3.0695	79	-	-	-

When analysing the variety with respect to plant height, it was found that there is a relationship between the two variables. From the values obtained it can be concluded that there are differences between the heights with respect to the varieties.Similar results were obtained by Grijalva-Contreras et al., (2009) and Santos et al; (2024) who state that the growth rate of olive trees is quite slow compared to other trees. On average, an adult olive tree can grow between 5 and 20 centimetres per year, although this depends on various factors such as climatic conditions, soil type, water availability and the amount of sun it receives. The height of these trees varies according to different factors.

On average, an adult olive tree can reach a height of between 5 and 10 metres, but there are exceptional cases where olive trees have been recorded with heights of up to 20 metres.

The height of an olive tree will depend on several factors such as the type of olive tree, its age, the climate in which it is located and the cultivation techniques used. In general, young olive trees have a lower height, while as they grow they can reach their maximum size.

In addition to the overall height of the tree, it is important to note that the height of the olive tree canopy can also be relevant depending on the type of crop being grown. Some crops require the canopy to be at accessible heights in order to harvest the fruit, while in other cases the canopy is allowed to grow more freely.

The "olive" is cultivated for the production of oil and/or table olives, with a high genetic diversity (Dridi et al., 2018; León et al; 2021). Therefore, the olive industry requires cultivars with adaptive resistance, especially in relation to those areas exposed to waterlogging as an effect of climate change.

4.1. Proposal for measures to mitigate crop damage caused by waterlogging

Disaster events associated with rain or the lack of it, regardless of today's considerations about the repercussions of climate change in terms of increased intensity and frequency, have an impact of damage and losses that are not only explained by the amount of water and the orographic conditions of the territory, but also by the elements of production, infrastructure, people and goods exposed, in a development dynamic with multiple environmental, economic and social factors that make us vulnerable.In this sense, it is possible to improve the resistance of plants to this adversity, such as grafting varieties onto more tolerant rootstocks in the case of fruit trees. Likewise, mycorrhization has shown good results in improving leaf retention and nutrient absorption in waterlogged plants.

Also, foliar application of nutrients and phytohormones prolongs crop tolerance to waterlogged conditions. In general, planting crops that are not tolerant to root hypoxia near rivers, water bodies and low valley sites should be avoided, relying instead on functional drainage or ploughing with deep subsoilers on well-levelled land. Another possibility is to plant the plants on ridges or to install deep ditches between the rows of plants (e.g. in papaya or banana) on sites prone to flooding.
Extreme events due to waterlogging will continue to increase, also in places where they were not expected before, and a multi-faceted approach is therefore necessary. In general, to try to improve the adaptation of the different olive varieties to waterlogging conditions, the following measures are proposed:

Creation of drainage systems:

Install drainage systems, such as drainage ditches, perforated pipes or soakaways. These systems will help divert excess water and keep the soil in a healthy condition for plants.

Figure 17 Drainage systems (Ditches).

The system of drainage channels will be complemented by a larger system of drains, furrows, ditches and collector channels that channel all the water from larger extensions until it is discharged into a receptor (lake). Consideration will be given to the preparation of a contour plan in order to determine the inflows and outflows of water and the most appropriate placement of the drains.The depth of drainage depends on the height of the water table, the expected maximum height of water during flooding, as well as the type of crop.Selection of species and varieties that are more resistant to flooding (flood-resistant crops) and maintenance of a good phytosanitary status of the plantations.Selection of species, considering flood-resistant species and maintaining a good phytosanitary status of the plantations. These measures are aimed at reducing the vulnerability of crops to flood damage.Damage caused by flooding is proportional to several factors, the most important of which are the species and the duration of flooding. In many cases, flooding does not cause the death of the crop, but rather reduces its growth and delays or shortens its production. It is in these cases that proper management can make an effective contribution to reducing damage.

Use of ground cover and minimum tillage

Ground covers are management practices to limit soil erosion caused by heavy rainfall events, they can also serve as a reservoir for useful fauna and help in pest control on the farm, however, they must be managed correctly so that they do not compete for water resources with the crop in times of greatest need.

In olive cultivation, in the last decade, the use of green cover has been introduced and extended as a cultural practice. Specifically, the increase in the area of olive groves with vegetation cover in this period was 2,93,158 ha (ESYRCE 2007 and 2017), 11% of the total olive grove.

Unlike other cultural practices, the use of ground covers in olive groves is associated with a multitude of positive environmental effects, such as the reduction of soil erosion, the increase of biodiversity, the increase of organic matter in the soil, the reduction of net greenhouse gas emissions and an improvement in the aesthetic quality of the landscape.

Creation of temporary controlled flood zones

One of the measures that can be applied to minimise the damage caused by floods is the creation of temporary controlled flood zones. These temporary controlled flood zones are very useful in those cases where the flood protection dyke system is overtopped by floodwaters, at which point the overtopped dykes become counterproductive.

Disease control

The incidence and severity of most of the diseases affecting olive groves are favoured in conditions of high humidity and high temperatures. Therefore, after prolonged periods of rain, a high incidence of diseases, both of the soil and of the aerial part of the plant, is to be expected.

Soil diseases:

Verticillium dahliae.

Verticillium is currently considered to be the most serious disease affecting olive groves, due to the significant damage it causes and, above all, its difficult control and rapid spread. Two different types of syndromes can be distinguished, one causing rapid drying of shoots and branches which can lead to the death of the tree, the other to slow decay,which is characterised by necrosis and mummification of the inflorescences that remain attached to the shoots. Root rot (Phytophthora spp., Fusarium spp., Armillaria spp., Phythium spp.) There are no specific effective control methods. The presence of these fungi is associated with high moisture content in the soil, so the same measures are advisable as for preventing root asphyxia.

Diseases of the aerial part:

Repilo (Spilocaea oleagina)

This fungus produces lesions on the leaf and fruit, causing them to fall. When it affects the fruit, it shrivels and deforms as growth stops in the affected area. Effective control can be achieved by foliar application of cupric fungicides at the right time and in the right quantity.Tuberculosis (Pseudomonas savastanoi pv. Savastanoi) This bacterium can cause the death of branches and shoots, as well as a progressive weakening of the tree, although it is very difficult for the plant to die. There is a loss of harvest as the number of fruits and their size decrease. Olives coming from affected branches have an unpleasant smell and a sour, bitter and rancid taste, giving oils with organoleptic conditions of lower quality. The fight against this disease must be essentially preventive, using varieties that are not very sensitive, limiting wounds and reducing epiphytic populations of the bacterium. In any case, it is recommended to treat in the event of a hailstorm or other type of accident that causes wounds to the plant, with products that have a bactericidal action.

CONCLUSIONS

The olive crop at Olive Land Farms has been affected by waterlogging, showing significant differences in plant height. Measures are proposed such as the creation of drainage systems, appropriate cultural work, selection of species and varieties with greater resistance to flooding, as well as the creation of controlled temporary flooding areas and disease control to mitigate the damage caused by waterlogging to the olive crop.

REFERENCES

• Antol, MN (2004). The Sophisticated Olive: The Complete Guide to Olive Cuisine. Garden City Park, NY: Square One Publishers. pp.37. ISBN 978-0-7570-0024-9. (registration required). "Kalamata olive.

• Aparicio, A, C., & Cordovilla, D. (2023). Olive (Olea europaea L.) and salt stress. Importance of growth regulators. University of Jaén. Faculty of Experimental Sciences.

• Arquero, O; Serrano, N (2020). Recommendations for olive groves with flooding problems. Available en: https://www.juntadeandalucia.es/agriculturaypesca/ifapa/servifapa/registro - servifapa/b174a396-bbd7-4daf-942f-e15eaefa34ae

• Balam, A, (2023). Characteristics of Arbequina olive trees. Retrieved from https://balam.es/

• Barranco, Navero D; Fernández, Escobar R; Rallo, Romero L. 2017. El cultivo del olivo 7th ed. Editorial Mundi-Prensa.

• Barranco, Navero D. (2008). Varieties and rootstocks. In: Barranco, D.; Fernández- Escobar, R.; Rallo, L. eds. El cultivo del olivo. Madrid, Mundi-Prensa. pp. 84-86.

• Bernal Martínez, J. 2017. Morphological and molecular characterisation of Italian varieties of olive (Olea europaea L.) installed in the introduction garden of INIA Las Brujas. PhD Thesis. Faculty of Agronomy, University of the Republic. Montevideo - Uruguay.

•Camacho Alonso, G. (2019). Physiological (trial in Carpio del Tajo-Toledo) and spectral (trial in Chozas de Canales-Toledo) effects of different doses of deficit irrigation in super-intensive olive groves (Arbequina). Polytechnic University of Madrid School of Agricultural, Food and Biosystems Engineering. Polytechnic University of Madrid. Available

at:https://oa.upm.es/56916/1/TFG_GEMA_MARIA_CAMACHO_ALONSO
.pdf

•Corral, M (2020). "Extra virgin olive oil: properties and benefits of our 'liquid gold' Editorial. El Español.com

•Díaz, C.; Moral, J.; Barranco, D.; Rallo, L. 2016. Genetic diversity and conservation of olive genetic resources. In: Ahuja, M. R.; Mohan Jain, S. eds. Genetic diversity. and erosion in plants, sustainable development and biodiversity. s.l., Springer. pp. 337-356.

•Doménech, 2020. Adaptation strategies for olive groves in the Serranos region (Valencia) in the face of climate change. Polytechnic University of Valencia. Escola técnica superior d'enginyeria agronòmica i del medi natural. Master's degree in agri-food and environmental economics.

•Dridi, J.; M. Fendri; C. M. Breton & M. Msallem. 2018. Characterization of olive progenies derived from a Tunisian breeding program by morphological traits and SSR markers. Scientia Horticulturae, 236, 127-136.

•References ESYRCE (2007): Crop Surface and Yield Survey. Ministerio de Medio Ambiente Medio Rural y Marino. https://www.mapama.gob.es/es/estadistica/temas/estadisticasagrarias/bol etin200 7_tcm30-122323.pd

•ESYRCE (2017): Survey on Crop Area and Yields. Ministry of Agriculture Fisheries y Food https://www.mapama.gob.es/es/estadistica/temas/estadisticasagrarias/bol etin201 7sm_tcm30-455983.pdf.

•Fischer, Gerhard (2021). Increased flooding caused by climate change will affect our crops. Revista Facultad Nacional de Agronomía Medellín, 74(3), 9619-9620. Epub September 26, 2021. Retrieved May 02, 2024. From. Available at :
http://www.scielo.org.co/scielo.php?script=sci_arttext&pid=S0304-

28472021000309619&lng=en&tlng=en.pdf.

• García Ruiz, R Torrús C and Calero G (2020). The olive grove and its adaptation to climate change. Available at: file:///H:/Grandes-cultivos/Articulos/302564-El-olivar-y- su-adaptacion-al-cambio-climatico.html

• Granitto , Ylenia (2016). Sustainable olive oil production helps mitigate climate change. Available in: Sustainable olive oil production helps mitigate climate change - Olive Oil Times.

• Grijalva-Contreras, R.L., MacíasDuarte, R., López-Carvajal, A., & Robles Contreras, F. (2009). Productivity of olive cultivars for oil (olea europea l.) under desert conditions in Sonora. Biotecnia, 11(2), https://doi.org/10.18633/bt.v1 1i2.60.

• Guerrero García, A. 2003. Olive grove implantation. New olive growing. 5th ed. rev. and expanded. Madrid, Mundi-Prensa. 304 p.

• León, L.; D. Casanova; J. Palma & J. González. 2021. Agromorphological characterisation of mother plants of the "olive" (Olea europaea (Oleaceae) germplasm bank in the Tacna region. Arnaldoa 28(3): 593-612 doi: http://doi.org/10.22497/arnaldoa.283.28307

• Lifeder (2023). Olive tree. Retrieved from: https://www.lifeder.com/olivo-olea- europaea/.

• Lorite IJ, Gabaldón-Leal C, Santos C, Cruz-Blanco M, León L, Porras R, Belaj A, de la Rosa R. 2019. Impact of climate change on Andalusian agriculture: olive groves. Institute for Agricultural and Fisheries Research and Training. Ministry of Agriculture, Fisheries and Rural Development. Junta de Andalucía.

• Penco, JM, 2019. Olivar y Cambio Climático: Impacto del Cultivo del Olivo sobre el Cambio Climático. Effect of Climate Change on the stability of the olive oil market in Spain. Olive Grove and Climate Change Conference. Beja Portugal.

•Rossi, L. (2023). Survey of Florida olives for small-scale commercial production. Rev. Oleoteca. Available at :https://www.oleorevista.com/texto-diario/mostrar/4202881/uf-ifas-prueba-olives-florida-determine-possible-commercial-industry

•Salgado, C A ; Avilés C D;. Martín, A D; Otero Cabeza D; Prieto, Ll; González,OJ; Soler, G V(2019). Guidelines for adaptation to flood risk: agricultural and livestock farms. Ministry for Ecological Transition General Technical Secretariat Publications Centre. Catálogo de Publicaciones de la Administración General del Estado: http//publicacionesoficiales.boe.es NIPO: 638-18-028-4. Available   en:
 file:///D:/Olivos/guia-adaptacion-al-risgo-risgo-inundacion- explotaciones-agricolas-ganaderas_tcm30-503727.pdf

•Santos E, Orestes, Valdés Reinoso, R. H., & Castillo Edua, B. R. (2024). Olive cultivation in Florida, an alternative for climate change adaptation. Avances, 26(1), 72-90.
http://avances.pinar.cu/index.php/publicaciones/article/view/805

•Savé, R. (2016). Measures for adaptation to climate change in olive groves. Olive groves and climate change. Jornada Olivar y Cambio Climático. Ministry of Agriculture, Fisheries and Food.

•Oleas G, (2023). Cuanto tarda en crecer un olivo en el mediterráneo, Available at https://www.ginartoleas.com

•Tapia C, F; Ibacache, F; Astorga, M. (2002). Manual of olive cultivation. Climate and Soil Requirements. Available at:
file://C:/Users/Desktop/Olivos/NR30540.pdf

•Terrasa, D. (2019). Florida: physical geography. The Guide, Geography. Available at: https://geografia.laguia2000.com/geografia-regional/florida-geografia-fisica.

•United States National Arboretum. Florida Hardiness Zones. St Johns River Water Management District. Accessed 11 March 2023.

Printed by Books on Demand GmbH, Norderstedt / Germany